Risk IT Framework

2nd Edition

ISACA.

About ISACA

For more than 50 years, ISACA® (*www.isaca.org*) has advanced the best talent, expertise and learning in technology. ISACA equips individuals with knowledge, credentials, education and community to progress their careers and transform their organizations, and enables enterprises to train and build quality teams. ISACA is a global professional association and learning organization that leverages the expertise of its 145,000 members who work in information security, governance, assurance, risk and privacy to drive innovation through technology. It has a presence in 188 countries, including more than 220 chapters worldwide.

Disclaimer

ISACA has designed and created *Risk IT Framework, 2nd Edition* (the "Work") primarily as an educational resource for professionals. ISACA makes no claim that use of any of the Work will assure a successful outcome. The Work should not be considered inclusive of all proper information, procedures and tests or exclusive of other information, procedures and tests that are reasonably directed to obtaining the same results. In determining the propriety of any specific information, procedure or test, professionals should apply their own professional judgment to the specific circumstances presented by the particular systems or information technology environment.

Reservation of Rights

ISACA

1700 E. Golf Road, Suite 400
Schaumburg, IL 60173, USA
Phone: +1.847.660.5505
Fax: +1.847.253.1755
Contact us: https://support.isaca.org
Website: www.isaca.org

Participate in the ISACA Online Forums: https://engage.isaca.org/onlineforums

Twitter: http://twitter.com/ISACANews
LinkedIn: www.linkedin.com/company/isaca
Facebook: www.facebook.com/ISACAGlobal
Instagram: www.instagram.com/isacanews/

Risk IT Framework, 2nd Edition
ISBN 978-1-60420-819-1

Acknowledgments

ISACA would like to recognize:

Lead Developer
Lisa Young, CISA, CISM, CISSP, Axio, USA

Expert Reviewers
Luis Alberto Capua, CRISC, CISM, Argentina
Tom Conkle, CISSP, Optic Cyber Solutions, USA
Andrew Foo, CISA, CRISC, CISM, CGEIT, CBCP, CCSK, CISSP, PMP, Dulwich College International, China
Sandra Fonseca, Ph.D., CISA, CRISC, CISM, CICA, Northcentral University, USA
Yalcin Gerek, CISA, CRISC, CGEIT, COBIT 5 Trainer, DASA DevOps Coach, ISO 20000LI, ISO 27001LA, ITIL Expert, PRINCE2, Resilia Practitioner, TAC, Turkey
Ahmad M. El Ghazouly, Ph.D., CISA, CRISC, CISM, PMI-ACP, AMBCI, BISL, PBA, PMP, PMI-RMP, TOGAF, PGESCo, Egypt
Demetri Gittens, CISA, CRISC, Central Bank of Trinidad and Tobago, Trinidad & Tobago
Ken Hendrie, CISA, CRISC, CISM, CGEIT, ISO27001 LI, ITIL, PRINCE2, IRAP, Cyconsol, Australia
John Hoffoss, CISA, CISSP, GCIH, CliftonLarsonAllen, USA
Mike Hughes, CISA, CRISC, CGEIT, MIoD, Prism RA, United Kingdom
John E. Jasinski, CISA, CRISC, CISM, CGEIT, CSX, COBIT 5 Assessor, COBIT and ITIL Accredited Instructor, AWS Practitioner, CCSK, Certified Scrum Master and Product Owner, ISO 20000, IT4IT, ITIL Expert, Lean IT, MOF, ServiceNow and RSA Archer Certified System Administrator, Six Sigma Blackbelt, TOGAF, USA
Jack Jones, CISA, CRISC, CISM, CISSP, RiskLens, USA
Linda Kostic, CISA, CISSP, Doctor of IT-Cybersecurity & Information Assurance, PRMIA Complete Course in Risk Management, George Washington University, Citi, USA
Jerry M. Kathingo, CRISC, CISM, Hatari Security, Kenya
Kamal Khan, CISA, CISSP, CITP, MBCS, United Kingdom
Shruti S. Kulkarni, CISA, CRISC, CCSK, CISSP, ITIL v3, Interpublic Group, United Kingdom
Jim Lipkis, Monoco Risk Analytics, Inc., USA
Tony Martin-Vegue, CISM, CISSP, Netflix, USA
Andre Pitkowski, CRISC, CGEIT, COBIT 5 Assessor, APIT Consultoria de Informatica Ltda, Brazil
Eduardo Oscar Ritegno, CISA, CRISC, Banco Nación, Argentina
Katsumi Sakagawa, CISA, CRISC, Japan
Gurvinder Pal Singh, CISA, CRISC, CISM, Qantas Airways, Australia
Darron Sun, CISA, CRISC, CISSP, CMA, CPA (Australia), CRMA, FIPA, Hong Kong Housing Society, China
Peter C. Tessin, CISA, CRISC, CISM, CGEIT, Discover Financial Services, USA
Alok Tuteja, CRISC, CGEIT, CIA, CISSP, BRS Ventures, United Arab Emirates
Ashish Vashishtha, CISA, CRISC, CISM, CIPT, CISSP, AWS Certified Cloud Practitioner, HITRUST CSF Practitioner, PROSCI Change Practitioner, AdventHealth, USA
Greet Volders, CGEIT, Voquals N.V., Belgium
Jonathan Waldo, CISA, CRISC, ITIL 4 Foundation, CH Robinson, USA
Larry G. Wlosinski, CISA, CRISC, CISM, CAP, CBCP, CCSP, CDP, CIPM, CISSP, ITIL v3, PMP, Coalfire-Federal, USA
Prometheus Yang, CISA, CRISC, CISM, CFE, Standard Chartered Bank, Hong Kong
Dušan Žikić, CISA, CRISC, CISM, CSX-P, Cybersecurity Audit, Cybersecurity Fundamentals, COBIT 5 Foundation, COBIT 2019 Foundation, COBIT 2019 Design and Implementation, ITIL (2011) Foundation, ITIL 4 Foundation, IBM Data Science,NIS Gazprom Neft, Serbia

Acknowledgments *(cont.)*

Risk IT Task Force
Steven Babb, CRISC, CGEIT, ITIL, MUFG Investor Services, United Kingdom
Urs Fischer, CISA, CRISC, CPA (Swiss), UBS Business Solutions AG, Switzerland
Jack Freund, Ph.D., CISA, CRISC, CISM, CISSP, RiskLens, USA
Apolonio Garcia, CRISC, Open FAIR, HealthGuard, USA
Jimmy Heschl, CISA, CISM, CGEIT, Red Bull, Austria
Gladys Rouissi, CISM, CRISC, ANC Wealth, Australia
James C. Samans, CISA, CRISC, CISM, CBCP, CISSP-ISSEP, CPP, PMP, American Institutes for Research, USA
Ekta Singh-Bushell, CISA, CGEIT, CISSP, CPA, Datatec, USA
Dirk Steuperaert, CISA, CRISC, CGEIT, IT In Balance, Belgium
Evan Wheeler, CRISC, IASO, Edelman Financial Engines, USA

Board of Directors
Brennan P. Baybeck, CISA, CRISC, CISM, CISSP, Vice President and Chief Information Security Officer for Customer Services, Oracle Corporation, USA, Chair
Rolf von Roessing, CISA, CISM, CGEIT, CISSP, FBCI, FORFA Consulting AG, Switzerland, Vice-Chair
Tracey Dedrick, Former Chief Risk Officer with Hudson City Bancorp, USA
Pam Nigro, CISA, CRISC, CGEIT, CRMA, Health Care Service Corporation, USA
R.V. Raghu, CISA, CRISC, Versatilist Consulting India Pvt. Ltd., India
Gabriela Reynaga, CISA, CRISC, COBIT 5 Foundation, GRCP, Holistics GRC, Mexico
Gregory Touhill, CISM, CISSP, AppGate Federal Group, USA
Asaf Weisberg, CISA, CRISC, CISM, CGEIT, introSight Ltd., Israel
Rob Clyde, CISM, Board Director, Titus and Executive Chair, White Cloud Security, USA, ISACA Board Chair, 2018-2019
Chris K. Dimitriadis, Ph.D., CISA, CRISC, CISM, Group Chief Executive Officer, INTRALOT, Greece, ISACA Board Chair, 2015-2017
Greg Grocholski, CISA, Saudi Basic Industries Corporation, USA, ISACA Board Chair, 2012-2013
David Samuelson, Chief Executive Officer, ISACA, USA

TABLE OF CONTENTS

LIST OF FIGURES

Chapter 1. Introducing the Risk IT Framework

Chapter 2. Risk IT Framework Principles

Chapter 3. Risk IT Framework Components and Alignment With COBIT

Chapter 4. Essentials of Risk Governance

Chapter 5. Essentials of Risk Management

Chapter 6. Essentials of Risk Assessment

Chapter 7. Risk Awareness, Reporting and Communication

Chapter 8. Essentials of Risk Response

Page intentionally left blank

Executive Summary

The concept of risk—when considered in the context of information technology and cybersecurity—references an extensive vocabulary, including threats and vulnerabilities, risk appetite, tolerance, impact, prioritization and response, among many other terms that are critical to the disciplines of risk governance, management and assessment of information technology.

Risk IT practitioners require complete, clear and distinct definitions of all of these terms in order to create a common language accessible to business partners across the enterprise. ISACA's risk IT framework develops the language of risk specifically in the context of information technology and cybersecurity, fosters open conversation about the countless facets of enterprise risk, codifies guidelines and practices that optimize risk, opportunity, security and business value, and helps practitioners build consensus regarding risk IT decisions at all enterprise levels. The *Risk IT Framework, 2nd Edition,* along with its companion volume, the *Risk IT Practitioner Guide, 2nd Edition*, facilitate collaboration among risk and IT practitioners, bringing the science of risk management to enterprise information and technology (I&T).

The profession of risk management is not a modern invention. Study of risk management originated in the 1600s when mathematicians Blaise Pascal and Pierre de Fermat exchanged letters about games of chance. Their correspondence is believed to have launched modern probability theory[1]—and ultimately, modern quantitative risk analysis. Yet in the context of today's information technology, information security, cybersecurity and cyberphysical systems,[2] risk management principles are not widely understood—especially their quantitative methods, which dramatically enhance awareness of I&T risk in business and financial terms.

Cyberrisk, in particular, may not be well understood by key enterprise stakeholders—including board members and executives, who depend on technology to achieve strategic and operational objectives and, consequently, should be accountable for risk management. Without a clear understanding of I&T-related risk, senior executives have no frame of reference for prioritizing and managing it—even though I&T increasingly dominate the ecosystem(s) in and through which the enterprise operates.

Risk—defined as the combination of the likelihood of an event and its impact—affords opportunities for benefit (upside) or perils to success (downside). Risk and opportunity go together; indeed, to provide business value to stakeholders, enterprises must engage in various activities and initiatives (opportunities), all of which carry degrees of uncertainty and, therefore, risk. Managing risk and opportunity is a key strategic activity for enterprise success.

The *Risk IT Framework, 2nd Edition* concerns the entire spectrum of I&T risk—any business or mission risk related to the use of, or dependence on, information and communications technology (ICT),[3] operational technology (OT),[4] network or internet of things (IOT),[5] electronic data, and digital or electronic communications. The framework is founded on the core principle of serving stakeholders and enhancing business value through effective enterprise governance and management of all types of I&T-related risk. In this publication, information security, information assurance and cybersecurity are treated as subdomains of I&T-related risk.

The framework embodies and elaborates several key guiding principles for risk IT across the enterprise:

- Connect management of I&T-related risk to business or mission objectives.
- Align the management of I&T-related business and/or mission risk with enterprise risk management (ERM), whenever possible, if ERM is operating in the enterprise.

1 American Physical Society, "July 1654: Pascal's Letters to Fermat on the 'Problem of Points'," *APS News*, vol. 18, no. 7, July 2009, *www.aps.org/publications/apsnews/200907/physicshistory.cfm*
2 See US National Institute of Standards and Technology (NIST), "Cyber-Physical Systems," *www.nist.gov/el/cyber-physical-systems*
3 NIST, "Information and Communications Technology (ICT)," Computer Security Resource Center, *https://csrc.nist.gov/glossary/term/information_and_communications_technology*
4 NIST, "Operational Technology (OT)," Computer Security Resource Center, *https://csrc.nist.gov/glossary/term/Operational-technology*
5 Voas, J.; "Networks of 'Things'," NIST SP 800-183, July 2016, *https://csrc.nist.gov/publications/detail/sp/800-183/final*

- Balance the costs and benefits of managing I&T-related risk with other enterprise risk.
- Promote ethical and open communication regarding I&T-related risk.
- Establish the tone at the top while defining and enforcing personal accountability for operating within acceptable and well-defined tolerance levels.
- Integrate risk IT practice into routine activities and processes—discontinuous, point-in-time or incidental efforts are intrinsically inimical to risk IT methodology.
- Take a consistent approach that is standard, repeatable and aligned to strategy.

I&T-related risk is one component within the overall risk universe of the enterprise (**figure 1.1**). Other types of risk include strategic risk, environmental risk, market risk, credit risk, operational risk and compliance risk. Some enterprises categorize I&T-related risk under operational risk—for example, those in the financial industry, as defined in the Basel II framework.[6] Yet all types of risk—even strategic risk—can include elements of I&T risk, especially if I&T form the core of new business initiatives. The same association applies to credit risk, for example—poor cyberrisk management may lead to security breaches and/or compliance penalties and, in turn, to lower credit ratings.[7]

The *Risk IT Framework, 2nd Edition* explains I&T-related risk and enables practitioners to:

- Identify and address I&T-related risk broadly at the enterprise level—not solely within the IT department
- Integrate the management of traditional IT risk, information security and cyberrisk into overall ERM processes
- Facilitate comprehensive, holistic, risk-aware decision making at the enterprise level
- Guide the enterprise risk response whenever I&T-related risk exceeds tolerance

The *Risk IT Framework, 2nd Edition* is part of ISACA's broad portfolio of I&T-related risk and governance products; it provides a complete, standalone framework, yet is closely aligned to COBIT®, and incorporates many of the same principles to achieve success. Its companion publication, the *Risk IT Practitioner Guide, 2nd Edition*, also aligns with COBIT. Both publications assume that risk IT practitioners understand the basic concepts of the COBIT framework.[8]

[6] Basel Committee on Banking Supervision, *Basel II: International Convergence of Capital Measurement and Capital Standards: a Revised Framework*, 10 June 2004, *https://www.bis.org/publ/bcbs107.htm*

[7] O'Flaherty, K.; "Equifax Becomes First Firm To See Its Outlook Downgraded Due To A Cyber-Attack," Forbes, 28 May 2019, *https://www.forbes.com/sites/kateoflahertyuk/2019/05/28/equifax-becomes-first-firm-to-see-its-outlook-downgraded-due-to-a-cyber-attack/#209549335671*

[8] For additional guidance, see ISACA, *Getting Started with Risk Management*, USA, 2018, *https://www.isaca.org/bookstore/bookstore-wht_papers-digital/whpgsr*; and ISACA, *Risk IT Practitioner Guide*, 2nd Edition, USA, 2020 (forthcoming). Both publications adopt COBIT methodologies and include a range of practical examples.

Chapter 1
Introducing the Risk IT Framework

1.1 The Imperative for Risk IT

I&T-related risk is an existential condition of digitally enabled business—whether or not the enterprise identifies its sources or recognizes its potential consequences. Exposure to cyberthreats in particular—a specific type of I&T-related risk—increases as enterprises integrate technology and leverage information to create value. Cyberthreats can have devastating impacts if not identified and managed appropriately.

Within this context, the Risk IT framework offers a structured, systematic methodology that enables enterprises to:

- Identify current and emerging risk throughout the extended enterprise
- Develop appropriate operational capabilities to ensure that business processes continue operating through adverse events
- Leverage investments in compliance or internal control systems already in place to optimize I&T-related risk
- Recognize I&T-related risk that exceeds the scope of technical controls and IT-related tools and techniques to integrate into the enterprise risk management (ERM) program
- Raise awareness of the balance between the benefits of technology and external partners (on the one hand), and the potential impact of cyberthreats, internal control failures, and risk introduced by vendors, suppliers and partners (on the other hand)
- Promote risk awareness, accountability and responsibility throughout the enterprise
- Frame I&T-related risk within a business context to understand aggregate exposure in terms of enterprise value
- Focus internal and external risk management resources to maximize enterprise objectives

Risk IT aligns with major ERM frameworks, including the COSO ERM framework[9] and *ISO 31000 Risk Management*[10]; however, their implementation is not a prerequisite for adopting the Risk IT framework. Enterprises that adopt the Risk IT framework typically apply many common ERM principles in their foundational risk processes—regardless of the type of risk under management.

If ERM is already implemented in some form, it is important to build on the existing ERM program in order to:

- Increase stakeholder buy-in and adoption, leveraging existing concepts, terminology and consensus
- Save time and money in training and implementation
- Avoid discontinuity related to the substitution of a new IT, cybersecurity or cyberrisk management framework or terminology

Building on an existing ERM program is especially important when identified I&T-related risk has the potential to impact the overall business or mission—not just one part of the enterprise.

The Risk IT framework bridges the gap between traditional, generic risk management frameworks (e.g., COSO ERM and ISO 31000) and domain-specific frameworks like those in cybersecurity (e.g., the NIST Cybersecurity Framework[11]), information security (e.g., ISO 27005[12]) or project management (e.g., PMBOK®[13]). The Risk IT

[9] Committee of Sponsoring Organizations of the Treadway Commission (COSO), "Guidance on Enterprise Risk Management," *https://www.coso.org/Pages/erm.aspx*

[10] International Organization for Standardization (ISO®), *ISO 31000 Risk Management*, 2018, *www.iso.org/iso-31000-risk-management.html*

[11] NIST, *Framework for Improving Critical Infrastructure Cybersecurity*, Version 1.1, April 2018, *https://www.nist.gov/cyberframework/framework*

[12] ISO, *ISO/IEC 27005:2018 Information technology — Security techniques — Information security risk management*, July 2018, *https://www.iso.org/standard/75281.html*

[13] Project Management Institute, "PMBOK® Guide and Standards," *www.pmi.org/pmbok-guide-standards*

framework provides an end-to-end, comprehensive view of risk related to the use of I&T and thoroughly covers risk management—from the tone and culture at the top, to front-line practitioner and operational issues.

Applying the I&T-related risk management practices described in the Risk IT framework provides tangible business and/or mission benefits—fewer operational surprises and failures, increased information quality and reliability, greater stakeholder confidence, reduced regulatory concerns, and innovative applications supporting new business initiatives.

In summary, the Risk IT framework enables enterprises to understand and manage exposure to danger, harm or loss that is related to the use of, or dependence on, information and communications technology, electronic data, and digital or electronic communications.

1.2 Definitions and Terminology

The *Risk IT Framework, 2nd Edition* uses the following terms to describe key contexts, processes and activities:

- **Enterprise**—A group of individuals working together for a common purpose, typically within the context of a business organization, such as a corporation, partnership, limited liability company, government or public agency, charity, nonprofit or trust
- **Organization**—The structure or arrangement of interrelated components of an enterprise, defined by a particular scope
- **Business or mission**—The strategic purpose for which the organization exists. In the scope of risk IT, an enterprise typically sets strategic objectives—e.g., delivering a product or service, meeting sales targets and generating revenue. The purpose of a mission-driven organization may be similar to a business enterprise, but often operates to meet government, military or nonprofit objectives.
- **Governance**—The framework and system ensuring that:
  - Stakeholder needs, conditions and options are evaluated to determine balanced, agreed-on enterprise objectives
  - Strategic direction is set, and goals are prioritized and supported through appropriate, timely decision making
- **Risk**—The combination of the likelihood of an event and its impact
- **InformationSecurity** —The enterprise discipline that protects information against disclosure to unauthorized users (ensuring confidentiality), improper modification (ensuring integrity), and nonaccess, when required (ensuring availability)
- **Cybersecurity**—The enterprise discipline that protects information assets by addressing threats to information processed, stored and transported via internetworked information systems
- **Cyberrisk**—Exposure to danger, harm or loss related to the use of, or dependence on, information and communications technology, electronic data, and digital or electronic communications. Typically, the realization of cyberrisk involves unauthorized access and/or unauthorized use of information and communications technology.

In the *Risk IT Framework, 2nd Edition*, the term I&T broadly encompasses all information and related technology, digital and electronic ecosystems, and includes information security, cybersecurity and associated disciplines and processes. The term IT in this framework refers more narrowly to a function or department, internal or external, that provides technology support.

The Risk IT framework applies proven, generally accepted concepts from major generic industry standards and, at times, also develops key concepts from other I&T risk management frameworks. However, the terminology of the Risk IT framework may differ from that of other guidelines. For practitioners who are familiar with other frameworks—or perhaps have implemented another standard—both the *Risk IT Framework, 2nd Edition* and the *Risk IT Practitioner Guide, 2nd Edition* integrate and extend common industry risk management concepts and terminology, and map key structures from the Risk IT framework to their logical counterparts in the other (familiar) standards.

1.3 Purpose of the Risk IT Framework

In many enterprises, I&T have become central to daily operations, and increasingly constitute the core of overall business value. I&T-related risk should therefore be treated like any other key business risk—e.g., strategic risk, environmental risk, market risk, credit risk, operational and compliance risk—all of which fall under the highest risk category, failure to achieve strategic enterprise objectives.

In some enterprises, IT-related risk, information security risk and cyberrisk are considered subcategories of operational risk. Although other types of critical risk have long been integrated into enterprise decision-making processes, executives still tend to relegate I&T-related risk to the domain of technical specialists outside the boardroom. I&T-related risk pervades the entire organization, and thus demands an integrated risk management approach—not isolated, local or *ad hoc* treatments.

The Risk IT framework explains I&T-related risk and enables users to:

- Identify I&T-related risk that exceeds narrow technical judgment and thus requires holistic, enterprise-level consideration
- Integrate the management of I&T-related risk into overall ERM processes
- Evaluate I&T risk and response in the context of overall enterprise risk tolerance

1.4 Background

I&T risk often arises at critical nodes between (or among) interconnected environments, including points of access to the Internet. These interconnections are nevertheless vital to business and mission—and for that reason, they often entail the most acute information security and cybersecurity risk.

Generally, information security seeks to protect information by maintaining its confidentiality, integrity and availability (CIA), and by securing the assets in which it lives. Other factors in information security include nonrepudiation, privacy and sensitivity. Today, cybersecurity risk virtually always permeates other types of risk, because technology is often the vector—or path—through which cyberrisk is realized.

In formulating a business or operational strategy, an enterprise often decides explicitly to accept some level of risk to achieve its objectives. In COBIT, this practice is known as optimization, i.e., maintaining risk within tolerance to the risk appetite, which should be the goal of risk management. The Risk IT framework primarily focuses on resources and activities that reduce business impact from a realized risk, or reduce the likelihood (or probability) of a risk materializing that exceeds acceptable levels. The framework broadly facilitates management of the entire spectrum of I&T-related risk; however, as relevant subcategories, information security and cyberrisk examples may be used to show the interrelated nature of systems and processes.

The Risk IT framework primarily focuses on resources and activities that reduce business impact from a realized risk, or reduce the likelihood (or probability) of a risk materializing that exceeds acceptable levels.

The Risk IT framework is not a standard, but a framework, and references COBIT governance and management objectives and practices. Enterprises should tailor the guidance in the framework to suit their particular industry and business context.

1.5 Target Audience and Stakeholders

The Risk IT Framework, 2nd Edition is written for a wide audience, because risk management is an all-encompassing, strategic requirement in any enterprise. The target audience for the *Risk IT Framework* includes:

- Top executives and board members who set strategic direction and monitor risk at the enterprise level
- Managers of IT, OT and business departments who are responsible for day-to-day operational decision making and integration of risk management processes into daily work
- Risk management professionals who need specific I&T, information security, cybersecurity or cyberrisk guidance
- External stakeholders such as clients, regulators, suppliers and partners

I&T-related risk extends across the entire risk universe of the enterprise (**figure 1.1**). Other types of enterprise risk include strategic risk, environmental risk, market risk, credit risk, operational risk and compliance risk. In the financial industry (to take one example), I&T-related risk is often regarded as a subtype of operational risk, as defined in the Basel II framework.[14] However, strategic risk can encompass I&T-related risk, especially where I&T are the foundation of new business initiatives. The same applies to credit risk: Poor cyberrisk management practices can lead to lower credit ratings.[15] The Risk IT framework treats I&T-related risk as a continuum, fully coextensive with other major categories of risk—not as a narrow subtype of risk, hierarchically subordinate to (or dependent on) one or another parent category. The conceptual subordination of I&T-related risk to another type of risk—or its confinement to one narrow department or division of the enterprise—might diminish risk awareness and assessment, and lead to poor risk judgment and/or misrecognition of its authentically universal scope.

Figure 1.1—Scope of I&T-related Risk Relative to Other Major Categories of Risk

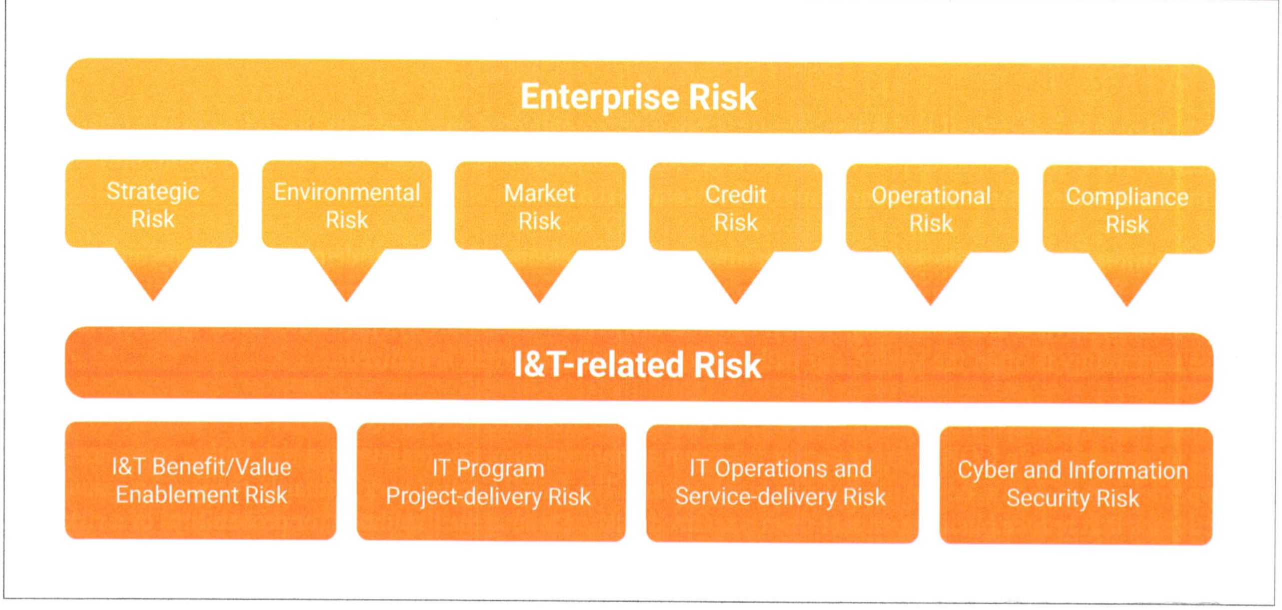

[14] *Op cit* Basel Committee on Banking Supervision
[15] *Op cit* O'Flaherty

Chapter 2

Risk IT Framework Principles

2.1 Introduction

The *Risk IT Framework, 2nd Edition* develops guiding principles for effective management of I&T-related risk—i.e., business risk related to the use of, or dependence on, information and communications technology, electronic data, and digital and electronic communications. Its principles are based on commonly accepted ERM principles, which have been applied to the domain of I&T. The Risk IT framework is designed to help enterprises apply the principles in practice.

IT, cybersecurity and information security transcend any one single, monolithic source or category of risk; they entail countless interrelated conditions, and reflect a multitude of specific, unique characteristics. They can involve any number of specialized technologies, threat actors, human errors, attack vectors, control failures and software vulnerabilities. It is especially important to note that cyberrisk and information security risk are not limited only to technology—many headline-grabbing risk events begin with human errors by real people.

Risk from IT, information security and cybersecurity is not the only I&T-related risk that warrants attention. Other operational risk types—including process failures and business or economic cycles—also need to be managed. Any I&T-related risk that jeopardizes the organization's business or mission should be managed from the perspective of overall enterprise objectives, and thus falls under the guiding principles of the Risk IT framework (**figure 2.1**):

- Connect management of I&T-related risk to business or mission objectives.
- Align the management of I&T-related business or mission risk with ERM when possible.
- Balance the costs and benefits of managing I&T-related risk.
- Promote ethical and open communication of all I&T-related risk.
- Establish the tone at the top while defining and enforcing personal accountability for operating within acceptable and well-defined tolerance levels.
- Use a consistent approach, integrated into daily activities, that is standard, repeatable and aligned to strategy.

Figure 2.1—Principles of Risk Management

Source: ISACA, *COBIT® 5 for Risk*, USA, 2013, fig. 15, *https://www.isaca.org/bookstore/cobit-5/wcb5rk*

2.2 Connect to Enterprise Business or Mission

Effective enterprise governance of I&T-related risk always connect to business or mission objectives:

- I&T-related risk, including cyberrisk, is treated as a business risk, not as a separate type of risk, and the approach to management is comprehensive and cross-functional.
- Governance of I&T-related risk contribute to business or mission outcomes. I&T support the achievement of business objectives, and any associated risk is expressed in terms of the impact and probability it may have on business objectives or strategy. Analysis of I&T-related risk considers the connection between business processes and supporting I&T assets, applications or infrastructure, and/or third-party dependencies.
- I&T-related risk management, including practices in information security and cybersecurity, all strive to advance the business or mission, rather limit or inhibit it.

2.3 Align With Enterprise Risk Management

Effective enterprise governance of I&T-related risk aligns its management with overall ERM:

- Business or mission objectives and risk appetite are clearly defined.
- Enterprise decision-making processes consider the full range of potential consequences and opportunities from I&T-related risk.
- The defined and stated risk appetite reflects the enterprise risk management policy and tone at the top, and influences the culture of the enterprise.
- I&T-related risk assessment is coordinated and consolidated across the enterprise (including, e.g., across information security and cybersecurity).

2.4 Balance Costs and Benefits

Effective enterprise governance of I&T-related risk balance its costs and benefits:

- I&T-related risk is prioritized and addressed in line with risk appetite and risk tolerance.
- Risk responses are implemented on the basis of cost/benefit analysis, analysis of alternatives and prioritization of risk that has the greatest potential impact on enterprise objectives.
- Existing controls and risk response actions are leveraged to address risk as efficiently as possible.

2.5 Promote Ethical and Open Communication

Effective management of I&T-related risk promotes ethical and open communication:

- Open, accurate, timely and transparent information on I&T-related risk is freely exchanged and informs risk-related decisions.
- Risk culture and risk management methods are integrated across the enterprise.
- Technical findings are translated into relevant and understandable business and financial terms.
- Information about an incident and the associated response is communicated openly to stakeholders, government and/or regulatory authorities, customers, and (as necessary) the public.

2.6 Establish Tone at the Top and Accountability

Effective management of I&T-related risk establishes an engaged tone from the top while defining and enforcing personal accountability for operating within acceptable and well-defined tolerance levels:

- Business owners, the board of directors and executive leadership are engaged in risk management.
- There is clear accountability and assignment of risk ownership.
- Risk assumptions are understood and supported by appropriate business leaders and are clearly stated in documentation of risk appetite, tolerances, culture, policies and guidelines for enforcement.
- Risk management performance is measured and integrated into the performance management of those accountable and responsible.
- Risk-aware culture and personal responsibility are promoted.
- Risk-informed decisions are made at the right level in the organization, by authorized individuals, in line with tolerances.
- Risk management practices are appropriately prioritized and embedded into enterprise decision making.

2.7 Use a Consistent Approach Aligned to Strategy

Effective management of I&T risk promotes continuous improvement and is part of daily activities:

- The dynamic nature of risk requires the enterprise to prepare by giving advance consideration to changes:
  - In the organization itself (mergers and acquisitions)
  - In the risk landscape
  - In applicable laws and regulations
  - In information and technology, as they evolve
  - In the industry at large
- Risk assessment methods, scales of measurement and criteria are consistent across the enterprise, especially as applied to:
  - Identification of key processes and associated risk
  - Identification of impacts on objectives
  - Identification of triggers that indicate when risk is out of tolerance or when an update of the framework or components in the framework are required, etc.
  - Monitoring and testing of operating controls
  - Actions to prevent risk from materializing
  - Risk response (if adverse events occur)
  - Identification and, to the extent possible, mitigation of assessor bias in the quantitative risk measurement process

Page intentionally left blank

Chapter 3

Risk IT Framework Components and Alignment With COBIT

3.1 Introduction

The Risk IT framework is built on the principles laid out in chapter 3 and further developed in subsequent chapters. This chapter discusses the:

- Components of the Risk IT framework
- Alignment of COBIT with the Risk IT framework
- Application of the Risk IT framework independent of COBIT

Figure 3.1 illustrates the alignment of I&T-related risk management principles with COBIT objectives EDM03 *Ensured risk optimization* and APO12 *Managed risk*.

Figure 3.1—Alignment of I&T-related Risk Management Principles With COBIT Objectives EDM03 and APO12

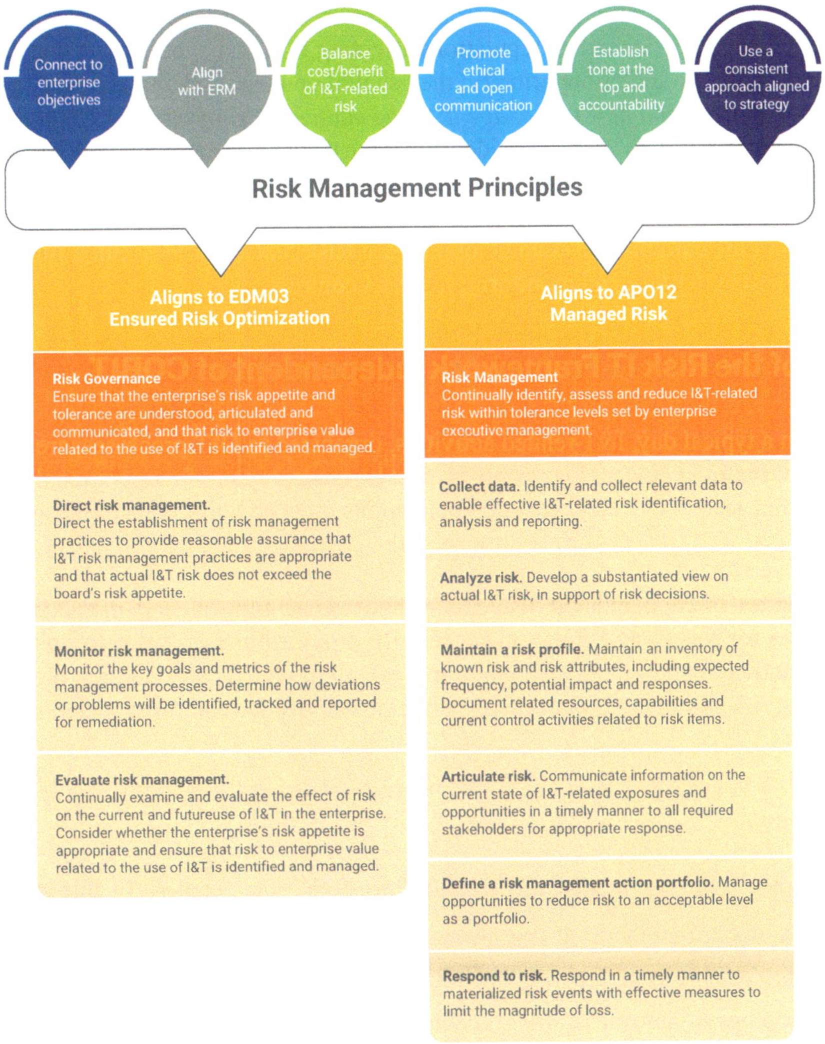

3.2 Components of the Risk IT Framework

The Risk IT framework is based on a set of guiding principles for effective management of I&T-related risk, and complements COBIT, a comprehensive framework for the governance and management of business-driven I&T solutions and services. The Risk IT framework enables enterprises to identify, govern and manage I&T-related risk.

Various different frameworks, techniques or methods can help enterprises establish and maintain capabilities for managing risk efficiently and effectively. Whether risk is managed holistically (as in ERM), or locally as a single type or category, (e.g., noncompliance, cybersecurity or information security risk), the underlying principles of risk management apply.

3.3 Alignment of COBIT With the Risk IT Framework

COBIT governance and management objectives and practices assist the enterprise in managing I&T processes, activities and services, whether internal or external to the enterprise, and help frame the overall enterprise engagement with I&T-related risk.

Internal enterprise events and activities can include operational or cyberincidents, project failures, business or technology strategy changes, evolutions, and mergers or acquisitions. External events can include changes in market conditions, competition, technology advances, regulations affecting I&T and cyberthreats that may arise from these events. All these factors entail risk and/or opportunity; they should be identified and assessed, and responses need to be developed. *The risk dimension and how to manage it, is the main subject of the Risk IT framework.* When opportunities for I&T-enabled business change are identified, the COBIT framework—specifically objectives EDM03 *Ensured risk optimization* and APO12 *Managed risk*—can prescribe practices and activities that enable enterprise objectives to be met. Risk management encompasses the sum total of activities and culture that create and preserve value in pursuit of the enterprise's strategic objectives. Risk management is not a function or a department, nor is it only limited to putting in place and monitoring internal controls.

3.4 Application of the Risk IT Framework Independent of COBIT

In a typical enterprise on a typical day, I&T-related activities, organized in I&T processes, are deployed. Events occur on a nonstop basis: Important technology choices must be made, repairs for operational incidents must be applied, software problems need to be addressed and applications must be built. Each of these events carries both risk and opportunity.

Risk reflects the combination of the likelihood of events and the impact they have on the enterprise. Risk therefore reflects both opportunities for benefit and threats to success. To provide business value to stakeholders, enterprises must engage in various activities and initiatives (opportunities), all of which carry degrees of uncertainty and, therefore, risk. Managing risk and opportunity is a key strategic activity for enterprise success.

Risk reflects the combination of the likelihood of events and the impact they have on the enterprise. Risk therefore reflects both opportunities for benefit and threats to success.

I&T can play different roles in the risk-opportunity relationship, both as an enabler of value and an impediment:

- **Value enabler**—New business initiatives almost always depend on some involvement of I&T. In this role, I&T can:
  - Enable successful I&T projects, support new initiatives and, thus, create value
  - Apply new technology in innovative ways, enable new business initiatives and create value
  - Protect assets and resources from threats that may impact delivery of products and services

- **Value impediment**—An array of negative consequences may result from I&T-related activities and processes:
  - I&T-enabled business projects or investments often fail to deliver expected results, so value is not delivered.
  - The enterprise may fail to identify or capture opportunities for new business initiatives arising from new technology.
  - I&T can fail to prevent or detect conditions or cyberthreats that can cause mild to serious operational disruption—e.g., system or network outages for short or extended durations; loss, disclosure or corruption of information.

How can an enterprise respond in practice? Ideally, the enterprise embeds both risk-aware and opportunity-aware thinking in the evaluation and monitoring of all I&T initiatives—not only those requiring involvement of the IT department or support function. For example, when an important investment in infrastructure is proposed, the enterprise should consider the following factors in its decision making:

- Risk associated with the investment, e.g., project risk
- Benefits of the new initiative in terms of risk reduction
- Business benefits of the resulting new I&T infrastructure
- Opportunities associated with the new I&T assets

When new technology emerges, the enterprise should consider the following criteria in determining whether to adopt the technology:

- Impact of adopting the technology (support, reliability, ease of integration)
- Risk associated with operating the new technology (e.g., security, reliability)
- Consequences of not adopting the new technology (e.g., obsolescence and lagging behind competitors)
- Business benefits of the new technology (e.g., support for new business initiatives, effectiveness and efficiency gains)

After the enterprise completes its initial assessment of the risk and/or opportunities, it should determine how to respond to them. A good risk analysis methodology reflects the guidelines described in this and other ISACA publications, and identifies the risk decisions to be made. Then, sound risk management and value management practices can be applied, enabling informed decision making.

Page intentionally left blank

Chapter 4

Essentials of Risk Governance

4.1 Introduction

This chapter discusses the essential components of risk governance. Although they are discussed briefly, more information and practical guidance can be found in COBIT. The topics covered here include:

- Risk appetite, risk tolerance and risk capacity
- Stakeholders for I&T-related risk management
- Risk culture

4.2 Risk Appetite, Risk Tolerance and Risk Capacity

In formulating strategies and/or operating plans, an enterprise must decide to take on some level of risk to achieve its objectives. An amount or magnitude of risk is generally expressed as risk appetite and risk tolerance. Although these terms are used frequently, the potential for misunderstanding is high. Some people use the concepts interchangeably, others see a clear difference. The Risk IT framework definitions are compatible with COSO ERM[16] and ISO 31000[17] definitions:

- **Risk appetite**—The broad-based amount of risk an enterprise or other entity is willing to accept in pursuit of its mission (or vision)
- **Risk tolerance**—The acceptable range relative to the achievement of a given objective (best when quantified in terms of in the same unit measure as the related objective)

Risk Appetite

Risk appetite reflects the amount of risk an entity is prepared to accept in order to achieve its objectives. When considering risk appetite levels for the enterprise, three major factors are important:

- The objective capacity of the enterprise to absorb loss—e.g., financial loss or damage to reputation
- The (management) culture or predisposition towards risk taking—e.g., cautious or aggressive. What amount or magnitude of loss will the enterprise accept to pursue its strategy or objectives?
- The nature of the business and the type of risk involved—e.g., the failure of a conveyor belt in a candy factory vs. the failure of a flight-control system on a commercial airliner

Risk appetite is different in each enterprise—there is no absolute norm or standard of what constitutes acceptable and unacceptable risk.

Statements of risk appetite are often broad, and tend to speak of risk hypothetically or generally—e.g., "the enterprise will not accept the risk of noncompliance," or "the organization will not accept fraud risk"—rather than express risk concretely in quantifiable terms. Although such representations of risk appetite are common, they are very difficult to cascade down through the organization as management directives: Absolute prohibitions on risk are impossible to maintain and therefore impractical. Under such a prohibition against risk, every control deficiency would be fixed, and every business endeavor with risk would be declined. In practice, this approach is not a productive or efficient use of resources. Instead, enterprises should attempt to determine a loss amount that is

[16] *Op cit* Committee of Sponsoring Organizations of the Treadway Commission (COSO)
[17] *Op cit* ISO

acceptable and manage to that amount. An example of a practical, concrete, quantified risk appetite statement might be:

Although the enterprise desires to have no appetite for I&T risk, it recognizes that this is impractical in the achievement of its objectives. Therefore, the enterprise will remediate loss scenarios whereby aggregate losses of $1 million or more are at risk.

Large enterprises may find it useful to have a version of this statement for each line of business. An enterprisewide appetite statement should reflect (or aggregate) all the line-of-business statements.

Every enterprise must define its own risk appetite levels and review them on a regular basis. This definition of risk appetite should align with the overall risk culture that the enterprise wants to express (i.e., ranging from very risk averse to risk taking/opportunity seeking). Although there is no universal right or wrong, risk appetite needs to be defined, well understood and communicated. Risk appetite and risk tolerance should be applied not only to risk assessments but also to all I&T-related decision making.

Risk Tolerance

Risk tolerance reflects a range of acceptable deviation from the level set by the risk appetite and business objectives—for example:

Standards require projects to be completed within estimated budgets and time frames, but overruns of 10 percent of budget or 20 percent of time are tolerated.

Guidance regarding Risk Appetite and Risk Tolerance

The following guidance applies to risk appetite and risk tolerance:

- Risk appetite and tolerance are defined at the enterprise level, reviewed and/or influenced by the board of directors, and reflected in strategy and policies set by executives. At lower (tactical) levels of the enterprise—or perhaps within certain enterprise entities or subsidiaries—exceptions can be tolerated (or different thresholds defined), as long as overall exposure at the enterprise level does not exceed the determined risk appetite. Any business initiative includes a risk component, so management should have discretion to pursue new opportunities in the context of risk. Enterprises that have a conservative risk appetite and tolerance policies could lack the agility or innovation to exploit new business opportunities. Conversely, risk appetite and tolerance policies can be dictated by legal, regulatory or industry requirements, and it may be appropriate to have no risk tolerance for failure to meet such mandates.
- Risk appetite and tolerances are defined, reviewed and updated periodically (as determined by the enterprise), and clearly communicated to all stakeholders. Risk exception processes are clearly defined and communicated.
- New market conditions, changing risk landscape, revised strategy and many other factors require the enterprise to reassess its risk portfolio and reconfirm its risk appetite at regular intervals, triggering risk policy reviews. In this respect, the enterprise should understand that risk management can provide value to the enterprise by allowing it to pursue risk-inclusive strategies and optimize allocation of resources.
- The costs of risk response or the business impact of risk may exceed the capabilities/resources of an enterprise, thus forcing higher tolerance for one or more risk conditions. For example, if a regulation says that sensitive data at rest must be encrypted—but no feasible encryption solution exists, or the cost of implementing a solution is prohibitive—the enterprise may choose to accept the risk associated with regulatory noncompliance, which is an enterprise-appropriate decision informed by factual data.

Risk Capacity

The term risk capacity is sometimes used in discussions of risk appetite. Risk capacity is usually defined as the objective magnitude or amount of loss that an enterprise can tolerate without risking its continued existence. As such, it differs from risk appetite, which generally reflects a board or management decision regarding how much risk is desirable, as illustrated in **figure 4.1**.

Figure 4.1—Risk Capacity, Risk Appetite and Actual Risk

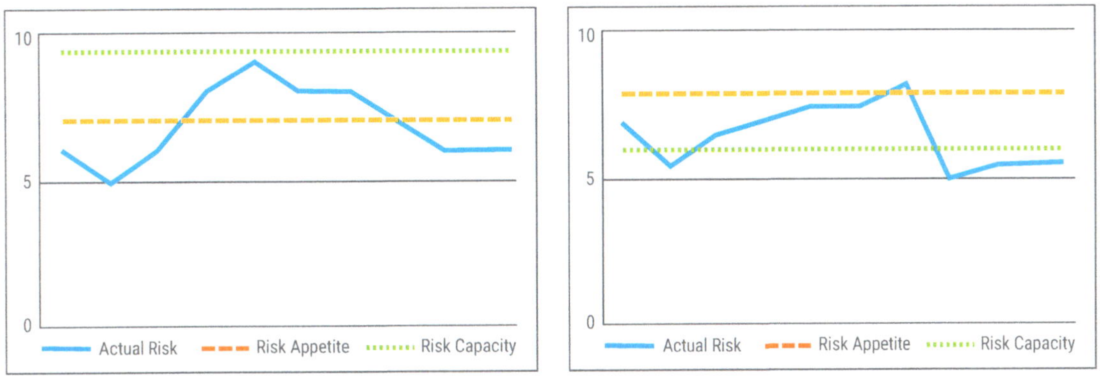

Source: ISACA, *COBIT® 5 for Risk*, USA, 2013, fig. 68, *https://www.isaca.org/bookstore/cobit-5/wcb5rk*

- The left diagram shows a relatively sustainable situation in which risk appetite is lower than risk capacity, and actual risk exceeds risk appetite in several situations, but always remains below the risk capacity.
- The right diagram shows a rather unsustainable situation, where risk appetite is defined by management at a level beyond risk capacity. Management is prepared to accept risk well over the objective capacity to absorb loss. As a result, actual risk routinely exceeds risk capacity, despite remaining below the risk appetite level most of the time.

4.3 Stakeholders for I&T Risk Management

Across enterprises, the stakeholders for I&T risk management often differ. Assignment of responsibility and accountability for I&T risk management varies widely, depending on the industry and type of enterprise. For example, in many financial institutions, the chief risk officer (CRO) is relegated to the role of oversight (or second line of defense[18]), while the business lines take primary (first line) responsibility and sometimes even accountability for risk decisions. In other commercial enterprises, the chief information security officer (CISO) takes responsibility for information security risk management, while accountability resides with the chief information officer (CIO) or chief digital officer (CDO).

Assignment of responsibility and accountability for I&T risk management varies widely, depending on the industry and type of enterprise.

Because the roles in **figure 4.2** are implemented differently across enterprises, they do not correspond consistently to the same organizational units or functions. For that reason, each role is briefly described. All roles listed in **figure 4.2** are considered stakeholders for the management of I&T-related risk.

[18] Regarding the three-lines-of-defense model, see The Institute of Internal Auditors® (IIA®), *IIA Position Paper: The Three Lines of Defense in Effective Risk Management and Control*, USA, 2013, *https://na.theiia.org/standards-guidance/Public%20Documents/PP%20The%20Three%20Lines%20of%20Defense%20in%20Effective%20Risk%20Management%20and%20Control.pdf*

Figure 4.2—Stakeholders for I&T Risk Management

Role/Structure	Description
Board of Directors	Group of the most senior executives and/or nonexecutive directors accountable for governance and overall control of enterprise resources
Executive Committee	Group of senior executives appointed by the board to ensure that the board is involved in, and kept informed of, major decisions (The executive committee is accountable for managing the portfolios of I&T-enabled investments and I&T-related services and assets, ensuring that value is delivered, and managing risk. The committee is normally chaired by a board member.)
Chief Executive Officer (CEO)	Highest-ranking officer charged with the total management of the enterprise
Chief Financial Officer (CFO)	Most senior official accountable for all aspects of financial management, including financial risk and controls and reliable and accurate accounts
Chief Operating Officer (COO)	Most senior official accountable for operation of the enterprise
Chief Risk Officer (CRO)	Most senior official accountable for all aspects of risk management across the enterprise (An I&T risk officer function may be established to oversee I&T-related risk.)
Chief Information Officer (CIO)	Most senior official responsible for aligning IT and business strategies and accountable for planning, resourcing and managing delivery of I&T-related services and solutions.
Chief Technology Officer (CTO)	Most senior official tasked with technical aspects of I&T, including managing and monitoring decisions related to I&T services, solutions and infrastructures (This role may be assumed by the CIO.)
Chief Digital Officer (CDO)	Most senior official tasked with putting into practice the digital ambition of the enterprise or business unit (This role may be assumed by the CIO or another member of the executive committee.)
I&T Governance Board	Group of stakeholders and experts accountable for guiding I&T-related matters and decisions, including managing IT-enabled investments, delivering value and monitoring risk
Architecture Board	Group of stakeholders and experts accountable for guiding enterprise-related matters and decisions and for setting architectural policies and standards
Enterprise Risk Committee	Group of executives accountable for enterprise-level collaboration and consensus required to support enterprise risk management (ERM) activities and decisions (An I&T risk council may be established to consider I&T-related risk in more detail and advise the enterprise risk committee.)
Chief Information Security Officer (CISO)	Most senior official accountable for all aspects of security management across the enterprise
Business Process Owner	Individual accountable for performing processes and/or realizing process objectives, driving process improvement and approving process changes
Portfolio Manager	Individual responsible for guiding portfolio management, ensuring selection of correct programs and projects, managing and monitoring programs and projects for optimal value, and realizing long-term strategic objectives effectively and efficiently
Steering Committee (Programs/Projects)	Group of stakeholders and experts accountable for guiding programs and projects, including managing and monitoring plans, allocating resources, delivering benefits and value, and managing program and project risk

Figure 4.2—Stakeholders for I&T Risk Management *(cont.)*

Role/Structure	Description
Program Manager	Individual responsible for guiding a specific program (including articulating and following up on goals and objectives of the program) and managing risk and impact on the business
Project Manager	Individual responsible for guiding a specific project, including coordinating and delegating time, budget, resources and tasks across the project team
Project Management Office	Function responsible for supporting program and project managers and for gathering, assessing and reporting information about the conduct of programs and constituent projects
Data Management Function	Function responsible for supporting enterprise data assets across the data life cycle and managing data strategy, infrastructure and repositories
Head Human Resources	Most senior official accountable for planning and policies regarding human resources in the enterprise
Relationship Manager	Senior individual responsible for overseeing and managing the internal interface and communications between business and I&T functions
Head Architect	Senior individual accountable for the enterprise architecture process
Head Development	Senior individual accountable for the I&T-related solution development process
Head IT Operations	Senior individual accountable for IT operational environments and infrastructure
Head IT Administration	Senior individual accountable for I&T-related records and responsible for supporting I&T-related administrative matters
Service Manager	Individual who manages the development, implementation, evaluation and ongoing maintenance of new and existing products and services for a specific customer (user) or group of customers (users)
Information Security Manager	Individual who manages, designs, oversees and/or assesses an enterprise's information security
Business Continuity Manager	Individual who manages, designs, oversees and/or assesses an enterprise's business continuity capability, to ensure that the enterprise's critical functions continue to operate following disruptive events
Privacy Officer	Individual responsible for monitoring risk and business impact of privacy laws and for guiding and coordinating the implementation of policies and activities that ensure compliance with privacy directives (In some enterprises, the role may be referenced as the data protection officer.)
Legal Counsel	Function responsible for guidance on legal and regulatory matters
Compliance	Function responsible for all guidance on external compliance
Audit	Function responsible for provision of internal audits

Source: Adapted from ISACA, *COBIT® 2019 Framework: Governance and Management Objectives*, USA, 2019, Appendix B, *https://www.isaca.org/bookstore/bookstore-cobit_19-digital/wcb19fgm*

4.4 Risk Culture

Risk management maximizes the value that an enterprise generates, while avoiding losses that negatively impact its ability to achieve its mission—or even to continue operating. A risk-aware culture promotes open discussion of risk, and acceptable levels of risk are understood and maintained. A risk-aware culture begins at the top, with board members and business executives who set direction, communicate risk-aware decision making and reward effective risk management behaviors. Risk awareness also implies that all levels within an enterprise understand how and why the enterprise responds to adverse I&T-related events.

A risk-aware culture promotes open discussion of risk, and acceptable levels of risk are understood and maintained.

Risk culture is not easy to describe. It consists of a series of behaviors, as shown in **figure 4.3.**

Figure 4.3—Relevant Behaviors for Risk Governance and Management	
General Enterprise Behavior	
Has a risk- and compliance-aware culture throughout, including the proactive identification and escalation of risk	The enterprise defines an approach to risk management and risk appetite and establishes a policy of zero tolerance for noncompliance with legal and regulatory requirements.
Has defined policies that have been communicated and that drive behavior	All personnel understand and implement the requirements of the enterprise as defined in relevant policies.
Shows active receptivity towards raising issues and acknowledging negative outcomes	Whistle-blowers are regarded as positive contributors to the enterprise. The blame culture is avoided. Personnel understand the need for risk awareness and reporting of potential exposures.
Recognizes the value of risk	Personnel understand the importance of maintaining risk awareness and the value that managing risk adds to their roles.
Has transparent and participative culture	Communication is open and facts are not omitted, misrepresented or understated. The negative impact of hidden agendas is avoided.
Shows mutual respect	Stakeholders and risk assessors are encouraged to collaborate, respected as professionals and treated as experts in their roles.
Accepts ownership of risk	Risk practices are incorporated throughout the enterprise. Accountability is clearly assigned and accepted. I&T-related risk is owned by the enterprise and not viewed solely as the responsibility of the IT department or IT risk function.
Allows risk acceptance as a valid option	Management understands the consequences of risk acceptance. Impact is determined to be within the enterprise's risk appetite.
Risk Professional Behavior	
Shows effort to understand what risk is for each stakeholder and how it impacts their objectives	Risk professionals understand the business impact of risk, including competitive, operational, regulatory and compliance requirements. Although risk may be common across a given industry, each enterprise is unique in terms of how risk affects its objectives.
Creates awareness and understanding of risk policy	Alignment of risk capacity, risk appetite and enterprise policy leads to effective risk strategy.

Risk Professional Behavior	
Fosters collaboration and two-way communication during risk assessment	Risk assessment is fundamentally accurate and complete, and addresses stakeholder needs.
Defines risk appetite clearly and communicates in a timely fashion with relevant stakeholders	Stakeholders manage risk more effectively and there is appropriate alignment with organizational strategy and objectives.
Sets policies that reflect risk appetite and risk tolerance	Employees and management operate within risk tolerance. Business lines apply formal risk appetite and tolerance to daily practice. There is a clear process for proposing and making changes to risk appetite levels, with senior management consideration and approval.
Supports effective risk practice	Stakeholders understand risk from common portfolio view (product, process) and apply risk-based decision making to daily practice.
Uses KRIs effectively as an early warning	KRIs are associated with valid metrics and can be used as an indicator of process or control failure. KRI metrics are available and accessible for regular reporting and relate to objectives.
Acts promptly on the basis of risk indicators or events that fall outside of appetite and tolerance	Risk indicators are linked to the management risk response and remediation activities.
Management Behavior	
Sets direction and demonstrates visible and genuine support for risk practices	Quality risk management practices are maintained through genuine support from senior management.
Engages with all relevant stakeholders to agree on actions and follow up on action plans	The correct stakeholders are appropriately involved in ensuring timely resolution of issues and achievement of business plans.
Obtains genuine commitment and assigns resources for execution of actions	Personnel are empowered to execute actions required by risk management decisions.
Aligns policies and actions to risk appetite	Management makes appropriate risk decisions in complying with policies. Risk adjusted revenue aligns with management expectations.
Monitors risk and progress against action plans	Remediation plans are completed within expected business time frames
	and have a positive impact on enterprise objectives.
Reports risk trends to senior executives and board	The timely reporting of risk trends proactively manages risk and avoids lost opportunities.
Rewards effective risk management	Good risk practice is acknowledged. Employees' performance goals and reward structures stimulate effective risk management practices and appropriate execution of mitigation actions.
Source: Adapted from ISACA, *COBIT® 5 for Risk*, USA, 2013, fig. 26, *https://www.isaca.org/bookstore/cobit-5/wcb5rk*	

Risk culture includes:

- **Behavior toward taking risk**—What are the norms and attitudes towards risk-taking, identification of risk and analysis of risk?
- **Behavior toward policy**—Is policy something that exists but is not followed? Do policies drive behavior? Are policies easy to read, understand and follow?

- **Behavior toward negative outcomes**—How does the enterprise deal with negative outcomes, policy exceptions, loss events, cyberincidents, missed opportunities and incident investigations? Will it learn from them and try to adjust, or will blame be assigned without treating the root cause?

Symptoms of an inadequate or problematic risk culture include:

- Misalignment of actual risk appetite, stated tolerances and risk policies
- Failure to align risk policy with management direction and/or organizational norms regarding compliance with policy
- Existence of a blame culture. This type of culture should be avoided, because it inhibits relevant and efficient communication. In a blame culture, business units tend to point the finger at the IT department—or at each other—when projects are not delivered on time or do not meet expectations. In doing so, they fail to realize how the business unit's involvement up front affects project success. In extreme cases, the business unit may assign blame for failure to meet expectations that it never clearly communicated. Blame diminishes effective communication across units, further exacerbating delays. Executive leadership must identify and quickly rectify a blame culture to foster collaboration throughout the enterprise.

Chapter 5
Essentials of Risk Management

5.1 Introduction

This chapter introduces the essential components of an overarching risk management process.[19] Topics discussed here include:

- Setting the context and scoping risk management
- Understanding the risk management workflow

5.2 Setting the Context and Scoping Risk Management

Positioning risk to the enterprise within the context of its mission, strategy and objectives is the first step to ensure every process and procedure that is carried out on a daily basis meets the long-term business objectives of the enterprise and is in alignment with its risk posture. This is known as setting the context for risk management. Pairing a risk-based approach with a strategic view of the enterprise enables communication and clarification of which uncertainties, or risk, have the greatest potential to jeopardize enterprise targets, objectives and mission.

Positioning risk to the enterprise within the context of its mission, strategy and objectives is the first step to ensure every process and procedure that is carried out on a daily basis meets the long-term business objectives of the enterprise and is in alignment with its risk posture.

Risk management requires an enterprise to:

- Define the scope within which risk management steps apply
- Set criteria against which identified risk is assessed or evaluated

Scope should be determined within the context of enterprise objectives. Setting the context will help enterprises limit the scope of initial risk assessment—e.g., to one business function, such as accounting—and understand how that scope fits within the context of the overall enterprise.

Establishing the criteria against which identified risk is evaluated is also an important part of the overall risk management process. The development of risk appetite and risk tolerances can assist enterprises in quickly evaluating and understanding whether risk aligns with risk appetite, or requires further analysis or investigation.

[19] For additional guidance, see *op cit* ISACA, *Getting Started with Risk Management* and *Risk IT Practitioner Guide, 2nd Edition.*

5.3 Understanding the Risk Management Workflow

Figure 5.1 captures the major phases of the risk management workflow. The steps in the diagram are not necessarily performed sequentially. Each enterprise should develop a workflow that supports the most efficient and effective means to accomplish the tasks.

Figure 5.1—Risk Management Workflow

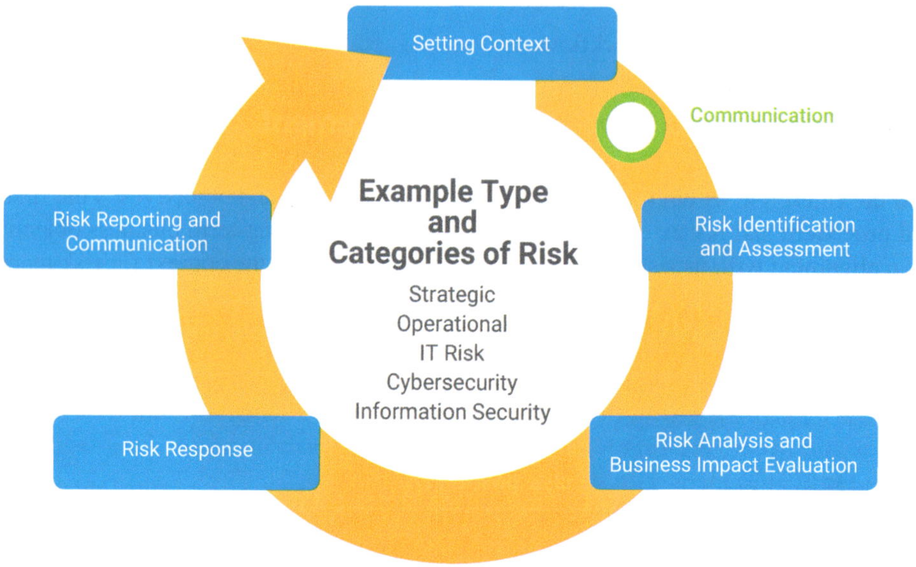

Source: Adapted from ISACA, *Getting Started With Risk Management*, USA, 2018, fig. 2, *https://www.isaca.org/bookstore/bookstore-wht_papers-digital/whpgsr*

Chapter 6
Essentials of Risk Assessment

6.1 Introduction

This chapter introduces essential components of the risk assessment process.[20] Topics discussed here include:

- Risk identification
- Risk analysis
- Evaluating the business impact(s) of identified risk
- I&T-related risk scenarios

6.2 Risk Identification

The risk identification process seeks to improve confidence that the enterprise recognizes and understands any risk with the potential to jeopardize its objectives.

Risk identification can occur in formal contexts (e.g., during brainstorming sessions or workshops) or informal settings (e.g., incidental discussion of issues in meetings or during office conversations). A brainstorming session usually starts with a list of items that keep participants up at night, including cyberthreats or other areas of concern. Often, issues that keep people awake contribute to risk—rather than contribute to the risk itself. For example, staff may worry about unpatched systems and often miscategorize them as risk.

The Risk IT framework seeks to identify loss event scenarios that may affect enterprise mission and strategic objectives. Their initial identification can occur in different contexts, or take different forms, including interviews, brainstorming activities, web self-reporting or surveys. Additional guidance is provided in the *Risk IT Practitioner Guide, 2nd Edition.*[21]

6.3 Risk Analysis

Risk analysis includes core approaches to enhance pragmatic insight, enterprise engagement and organizational transparency in the complex management of enterprise risk—especially I&T-related risk. Risk analysis is the process used to:

- Estimate the frequency and magnitude of a given risk scenario
- Identify and evaluate risk, its potential impact on the enterprise, and the likelihood (probability) that a particular event will occur

Risk assessment is slightly broader than risk analysis and includes the activities of ranking or prioritizing an identified risk according to defined enterprise risk thresholds, grouping like risk types together for mitigation, and documenting existing controls that provide mitigation.

6.4 Evaluating the Business Impact of Identified Risk

Meaningful risk assessments and risk-based decisions require I&T-related risk to be expressed in unambiguous, business- or mission-relevant terms. Effective risk management requires mutual understanding between IT and the

[20] For additional guidance, see *op cit* ISACA, *Getting Started with Risk Management* and *Risk IT Practitioner Guide, 2nd Edition.*
[21] *Op cit* ISACA, *Risk IT Practitioner Guide, 2nd Edition*

business regarding which risk needs to be managed and why. All stakeholders must be able to understand and express how I&T-related failures, compromises, mistakes or events can impact enterprise objectives and result in direct (i.e., financial) or indirect (i.e., data or information) loss (e.g., loss of sensitive customer information). Losses to the enterprise from I&T-related events can affect the enterprise's ability to deliver its key services and products.

Effective risk management requires mutual understanding between IT and the business regarding which risk needs to be managed and why.

The link between I&T risk scenarios and ultimate business or mission impact needs to be established to understand the effects of adverse events. Several techniques can help the enterprise describe I&T risk in business or mission terms. While the Risk IT framework requires I&T-related risk to be translated into or expressed in business-relevant terms, it does not prescribe any single method; several approaches are explored in The *Risk IT Practitioner Guide, 2nd Edition.*[22]

6.5 I&T Risk Scenarios

One of the challenges in I&T risk management is identifying the relevant risk in the context of everything that can possibly go wrong with I&T or in relation to I&T, especially given the pervasive presence throughout the enterprise.

One technique to overcome this challenge is development of risk scenarios, which bring insight and structure to the complex matter of I&T-related risk. After scenarios are developed, they are used during risk analysis, where frequency and business impact are estimated.

Risk scenarios can be derived via two mechanisms:

- **Top-down approach**—Mission strategy and business objectives form the basis for identifying and analyzing risk that is plausible and relevant to desired outcomes. If impact criteria are well aligned with the real value drivers of the enterprise, relevant risk scenarios can be developed.
- **Bottom-up approach**—Beginning with assets, systems or applications deemed important to the enterprise, a list of threats or generic loss scenarios is compiled. The resulting list is then used to define a set of concrete, customized scenarios that are applied to the enterprise context. The bottom-up approach is commonly used in cyberthreat and vulnerability assessments; however, it may limit visibility or obfuscate business impact, if its results are not considered in conjunction with the top-down approach.

The top-down and bottom-up approaches are complementary, and should be used together. A taxonomy of risk may help correlate their results, by providing a schema for classifying sources and categories of risk. The path from a cyberthreat (or area of concern) to a developed and documented risk requires the statement of risk to be decomposed into actionable components. The risk taxonomy provides a common language of discrete sources and categories, and helps practitioners communicate risk to stakeholders, ensuring that risk scenarios are relevant and linked to real business or mission risk.

After the set of risk scenarios is defined, it can be used for risk analysis, to assess frequency and impact of the scenario. An important component of this assessment are risk factors. Risk factors influence the frequency and/or business or mission impact of risk scenarios; risk factors can be of different types and are classified into two major categories:

- **Contextual factors (internal or external)**—The main difference is the degree of control that an enterprise has over the respective factors.
  - Internal contextual factors are, to a large extent, under the control of the enterprise, although they may not always be easy to change.
  - External contextual factors are, to a large extent, outside of the control of the enterprise.

[22] *Ibid.*

- **Capability factors (indicating ability to perform I&T-related activities)**—These factors are critical to successful outcomes in managing risk. Capability factors are embedded in many related ISACA tools, techniques, methods and frameworks that support an enterprise in defining and improving I&T and related processes needed to continue operating I&T-related activities. Capability factors help answer these questions:
    - I&T-related risk management capabilities—To what extent is the enterprise mature in performing risk management?
    - I&T-related business or mission capabilities (or value management)—How robustly do I&T-related capabilities support enterprise objectives while managing the risk that can jeopardize objectives?

An I&T risk scenario describes an I&T-related event that can lead to a business impact, when and if it occurs. For risk scenarios to be complete and usable for risk management and decision analysis, they should describe the following items, shown in **figure 6.1:**

- **Actor who generates the threat**—Actors can be internal or external, and they can be human or nonhuman.
    - Internal actors are within the enterprise—e.g., staff or contractors.
    - External actors include outsiders, competitors, regulators and the market.
    - Not every type of threat requires an actor—e.g., process failures or natural disasters.
- **Type of condition or nature of event**—Types of condition or event include: malicious, accidental, process failure, natural (i.e., *force majeure*), business cycle, etc.
- **Type of impact or outcome from the event**—Types of impact or outcome include: disclosure of information, interruption of systems, unintended modification or change, theft, destruction, etc. The events can reflect ineffective design (of systems, processes, etc.), ineffective execution of processes (e.g., change management procedures, acquisition procedures and/or project prioritization processes), effects of regulation, and inappropriate use. Impact also includes the cost of cleanup and remediation from the scenario.
- **Target asset or resource**—An asset is anything of value to the enterprise in the fulfillment of its mission or business strategy that can be adversely affected and lead to business or mission impact. A resource is anything that helps to achieve I&T-related goals. Assets and resources can be identical. For example, I&T hardware is an important resource (because all I&T-related applications use it) and simultaneously an asset (because it has a certain value to the enterprise). Assets/resources include:
    - **People**—e.g., employees, contractors, staffing providers and third parties
    - **I&T processes**—e.g., business and IT processes, data-flow diagrams or information flows
    - **Physical infrastructure**—e.g., facilities and equipment
    - **I&T infrastructure**—e.g., computing hardware, network infrastructure and middleware
    - **Other enterprise architecture components**, including:
        - Information
        - Applications

Figure 6.1—I&T-related Risk Scenario Development

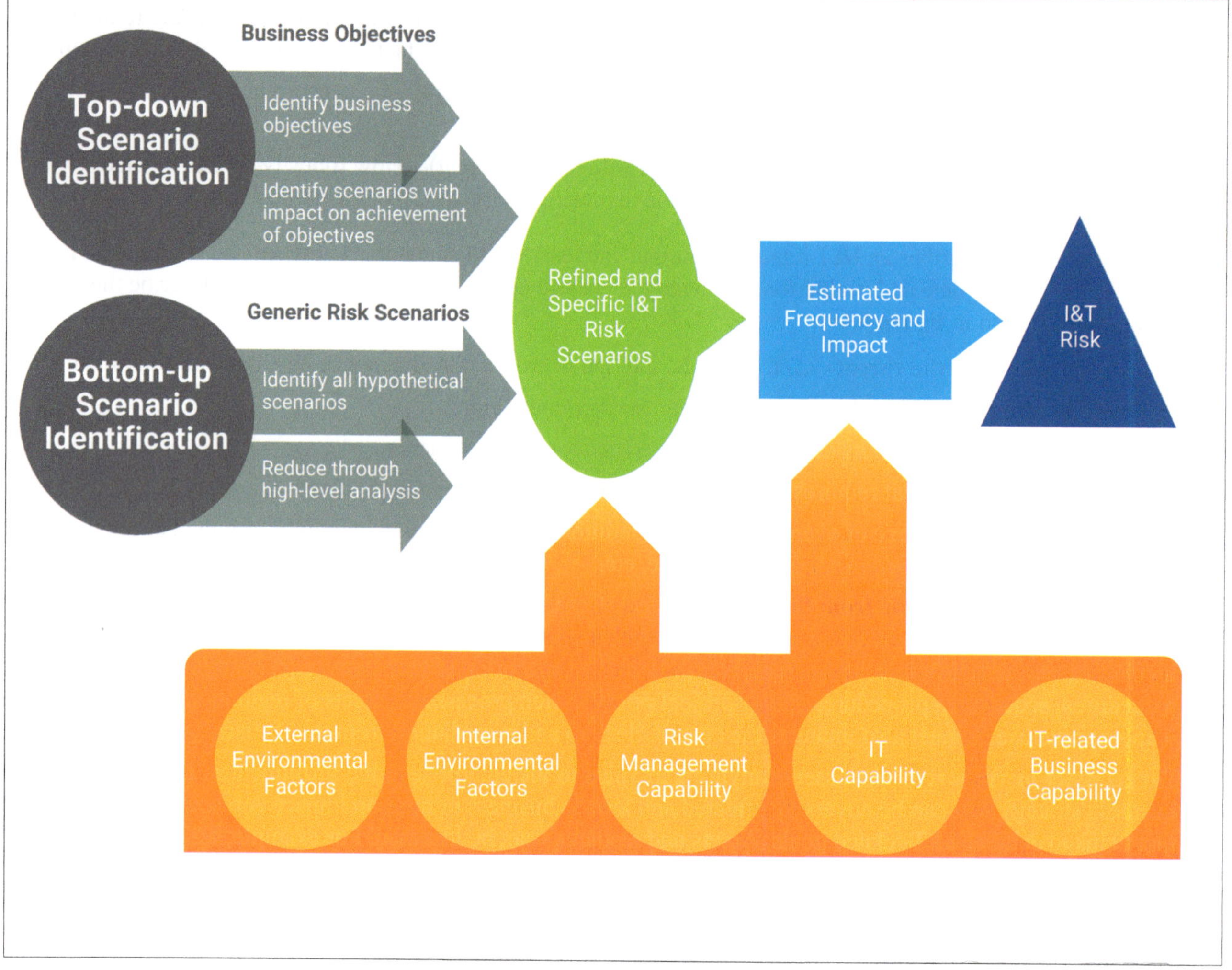

Some assets may be prioritized as critical, while others are regarded as noncritical (or only intermittently critical, at certain points in the business cycle). Critical resources may be targeted by a greater number of cyberattackers; therefore, the frequency of related scenarios will probably be higher. It takes skill, experience and thorough understanding of dependencies to distinguish between a critical asset and a noncritical asset.

Timing may also be relevant to some scenarios and is described as follows:

- **Duration of the event**—e.g., extended outage of a service or data center
- **Timing**—Does the event occur at a critical moment? Timing may further distinguish:
  - **Time lag between the event and the impact**—Is there an immediate consequence (e.g., network failure and immediate downtime) or a delayed impact over some extended duration (e.g., relating to obsolete IT architecture with cumulatively higher costs over several years)?

The risk scenario structure (**figure 6.2**) distinguishes loss events (events generating negative impact), vulnerabilities or vulnerability events (events contributing to the magnitude or frequency of loss events), and threat events (circumstances or events brought about by a threat actor that can trigger loss events).

- Actor/threat community
- Intent/motivation
- Threat event
- Asset
- Effect
- Timing

Figure 6.2—Risk Scenario/Loss Event Structure and Components

It is important to describe and understand the different components of the risk scenario so that appropriate actions can be taken. This is difficult to do with one large list of unprioritized conditions that may occur; therefore, practitioners prefer a focused, developed and nuanced list of relevant risk items that have been analyzed and prioritized by business impact. The risk register may be used to document and track risk that has been identified, analyzed and prioritized.

The *Risk IT Practitioner Guide, 2nd Edition*[23] includes further guidance for compiling relevant and manageable sets of I&T risk scenarios, and includes a starter set of example risk scenarios.

[23] *Ibid.*

Page intentionally left blank

Chapter 7
Risk Awareness, Reporting and Communication

7.1 Introduction

Risk awareness involves the acknowledgment that uncertainty, or risk, is an integral part of the business. This does not imply that all risk is to be avoided or eliminated, but rather that I&T-related risk should be:

- Identifiable
- Recognized
- Well understood and known
- Managed through application of appropriate resources

Risk reporting and communication are key parts of risk awareness. It is critical for decision makers and stakeholders (including boards of directors) to receive timely and accurate risk information on which they can act. People are often uncomfortable talking about risk; they tend to put off discussion of risk, because it involves contemplation of future uncertainty and, after all, might not actually materialize. Despite these subjective reactions, however, good communication regarding risk—i.e., *before* it is realized as an issue, incident or major crisis—is imperative.

It is critical for decision makers and stakeholders (including boards of directors) to receive timely and accurate risk information on which they can act.

7.2 Benefits of Risk Awareness and Communication

The benefits of open communication regarding I&T-related risk include the achievement of:

- Common understanding of actual exposure and potential impacts of a realized risk, enabling an appropriate and informed decision on risk response
- Transparency to all stakeholders regarding the potential level of risk exposure and risk management processes and capabilities in use

Poor communication regarding risk typically results in a:

- False sense of confidence regarding the actual degree of risk exposure
- Lack of well-understood direction for risk management from the top down
- Poor understanding on the part of stakeholders regarding level of risk exposure
- Perception that the enterprise is hiding known risk from stakeholders, regulators, investors or third parties (e.g., clients)
- Inability to respond in a timely manner to issues that can cause harm or loss
- Significant reputational damage or lowered expectations on the part of stakeholders when senior management is thought to be accountable, but fails to take corrective action, or does not adequately represent the corrective action to constituents

7.3 Risk Reporting and Communication

I&T-related risk communication covers a broad array of information flows. The Risk IT framework distinguishes the following major types of I&T-related risk communication, as shown in **figure 7.1**:

- **Expectations regarding risk management strategy, policies, procedures, awareness, training, etc.**—Enterprises should continuously communicate strategy and reinforce principles, etc., regarding the overall enterprise strategy for I&T-related risk. Clear and consistent communication of acknowledged risk drives all subsequent risk management efforts, raises awareness, and sets overall expectations for risk management behaviors.
- **Capability of current risk management**—Communicating enterprise risk management capability indicates how well the enterprise is managing risk and reducing exposure, facilitates transparency regarding any gaps in risk management competency, and is generally a key indicator of good risk management.
- **Status of identified risk under management**—Communications of risk status can include information from the following risk-related artifacts:
    - **Risk profile**—i.e., the overall portfolio of identified I&T-related risk to which the enterprise is exposed, including measures of each risk scenario in the portfolio
    - **Key risk indicators (KRIs)** to support management reporting on risk
    - **Event/loss data** regarding realized risk
    - **Root cause analysis** of realized loss events
    - **Mitigation options** (in terms of costs and benefits)

Figure 7.1—Components of I&T Risk Communication

To be effective, all information exchanged—regardless of type—should be clear, concise, complete, accurate, timely and understandable to all stakeholders. These criteria are especially important for information security, technology and cyberrisk. Jargon and technical terms regarding risk should be avoided. Extraneous or excessively detailed information hinders, rather than enables, a clear view of risk—particularly information regarding cyberthreats, vulnerabilities and events for which little factual evidence exists to indicate root cause(s) or the actual extent of any loss.

Critical time can elapse between identification of risk, its business or mission impact, and response activities. For example, a risk scenario may originate when an inadequate IT organization is set up. Its business impact is realized (eventually) in terms of inefficient I&T operations and service delivery. The scenario of IT project failure may result in eventual delays or failure to complete business initiatives. Communication is timely when it allows action to be taken at the appropriate moments to identify and treat the risk.

Information must be communicated at the right level of detail and adapted for the audience. In this process, aggregation must not hide root causes of risk. For example, a security officer needs technical I&T data on intrusions and viruses to deploy solutions. An I&T steering committee may not need this level of detail, but it does need aggregated information to decide on policy changes or additional budgets to treat the same risk.

Information must be communicated at the right level of detail and adapted for the audience.

Information must be available when needed by the appropriate audiences. Note that a risk register (including all documented risk) is not public information and should be properly protected against internal and external parties with no need to access it.

Communication does not always need to be formal, through written reports or messages. Timely face-to-face meetings between stakeholders are also important means to communicate I&T-related risk information.

7.4 Key Risk Indicators

Key risk indicators (KRIs) are metrics capable of showing that the enterprise is subject to—or has a high probability of being subject to—a risk that exceeds the defined risk appetite or tolerance. As their name implies, they are only indicators of risk and not direct measures of risk. It is important not to confuse risk measurement (and the corresponding assignment of risk ratings) with KRIs. They are specific to each enterprise, and their selection depends on many parameters in the internal and external environment, including the size and complexity of the enterprise, regulatory context (i.e., whether it operates in a highly regulated market), and strategy focus. Identification of KRIs should take into account the following steps (among others):

1. **Consider stakeholder needs when developing indicators of risk.** KRIs should be identified for relevant stakeholders based on information needs. Involving the right stakeholders in the selection of risk indicators also ensures greater buy-in and ownership.
2. **Iterate and improve indicators over time.** Take a balanced approach in selecting indicators that are forward looking, or *leading,* and backward looking, or *lagging.*

Leading indicators include pieces of data, information or capabilities that are in place to prevent events from occurring. Leading indicators may have upper and lower limits to help an enterprise understand when a condition requires attention before the risk is realized.

Lagging indicators include data, information or capabilities that are measured after an event or condition occurs—e.g., meeting a performance target or service level availability goal. Analyzing root causes from realized risk, failed controls or processes, and missed targets over time can help enterprises develop new indicators, trends or correlating conditions to gain insights.

An enterprise may develop an extensive set of metrics to serve as risk indicators; however, it is not generally feasible to maintain a large set of KRIs. By definition, *key* indicators are differentiated as being highly relevant and possessing a high probability of predicting or indicating risk outcomes.

The selection of the appropriate KRIs provide the following benefits to the enterprise:

- **Early warning, forward-looking signals** to the enterprise that a risk may be soon realized, enabling a proactive response before the risk becomes a loss
- **Backward-looking historical context** on risk that has been realized, further informing future risk responses, driving improvement and supporting documentation and analysis of trends
- **Feedback on risk appetite and tolerances** to facilitate improvements in risk management strategy and processes, and optimize risk governance and oversight

Common challenges or pitfalls associated with KRIs include:

- Undefined measurement objectives, absence of stated/expected outcomes, or lack of definitive questions that can be answered with data from KRIs
- Rote collection of data that is easy to obtain or already on hand, as opposed to data that correlates significantly to specific risk or risk types
- Lack of clear logical relationship between KRIs and specific risk or business objectives
- Excess of metrics without a clear measurement objective or purpose
- Cumbersome processes of aggregation
- Excess complexity in synthesizing and/or interpreting results from KRIs at an enterprise level

Because the internal and external environments are constantly changing, the risk environment is also highly dynamic, and the set of KRIs needs to be changed over time. Each KRI should be clearly related to the risk appetite and tolerance, so that trigger levels can be defined in support of appropriate, timely action.

Chapter 8
Essentials of Risk Response

8.1 Introduction

This chapter briefly discusses the essential components of risk response:

- Risk disposition
- Risk aggregation
- Risk response selection and prioritization

The following four risk dispositions help enterprises manage risk efficiently, focusing on risk with the greatest potential impact on objectives (should the risk materialize):

- Risk avoidance
- Risk mitigation
- Risk sharing or transfer
- Risk acceptance

The purpose of risk response is to bring risk in line with defined risk appetite in the wake of risk analysis. A response needs to be defined so that future residual risk (i.e., current risk after the risk response is defined and implemented) is, as much as possible (usually depending on available budget), maintained within risk tolerance limits. Management may decide to accept any risk, regardless of circumstances.

More information and practical guidance regarding risk response can be found in the *Risk IT Practitioner Guide, 2nd Edition*.[24]

8.2 Risk Avoidance

Risk avoidance entails exiting the activities or conditions that give rise to risk. Avoidance applies when no other risk response is adequate:

- No other cost-effective response can succeed in reducing the impact of the realized risk below defined threshold(s) for loss.
- The risk cannot be shared or transferred.
- The risk is deemed unacceptable by management.

Some I&T-related examples of risk avoidance include:

- Relocating a data center away from a region with significant natural hazards
- Declining to engage in a very large project when the business case shows notable risk of failure

8.3 Risk Mitigation

Mitigation reduces the frequency and/or impact of a risk. Common strategies for mitigation include:

- **Strengthening overall risk management practices**—Enterprises should consider assigning responsibility for risk identification and/or management to those closest to the activities or processes that generate the risk.

[24] *Ibid.*

- **Embedding risk awareness into regular workflows**—Enhancing risk awareness in the course of daily activities helps staff better understand and recognize risk-generating behaviors before an incident materializes.
- **Improving risk management processes and developing relevant tolerances**—Enterprises should seek opportunities to cascade and expand risk management from strategy to the front lines of the enterprise.
- **Automating triggers or alerts**—Automation generally affords the most advanced, timely indication when thresholds are out of tolerance.
- **Introducing controls**—Controls are intended to reduce the frequency or impact of realized risk. Various control techniques are discussed in the following sections.

8.4 Risk Sharing or Transfer

Sharing entails reducing risk frequency or impact by transferring a portion of the risk. Common techniques include:

- Obtaining insurance coverage for I&T-related events or cyberincidents
- Outsourcing I&T-related activities
- Sharing I&T-related project risk with a third-party provider through fixed price arrangements or shared investment arrangements

In neither concrete experience, nor in a more abstract legal sense, will these techniques relieve an enterprise of risk—however, they can leverage the skills of another party in managing the risk and, thus, reduce its financial impact, should an adverse event occur.

8.5 Risk Acceptance

Acceptance means that no action is taken relative to a particular risk, and loss is accepted when/if it occurs. This response is quite different from simply being ignorant of risk. Accepting risk assumes that the risk is known—i.e., an informed decision is made by management to accept it as such.

If an enterprise adopts a risk acceptance stance, it should carefully consider who can accept the risk—especially in the case of I&T-related risk—which should be accepted only by business management (and business process owners) in collaboration with (and supported by) the IT department or IT support function. Acceptance should be communicated to appropriate stakeholders, such as senior management and the board of directors, as necessary, and dictated by policy. Identification or mitigation of every risk may not be relevant or cost effective.

8.6 Risk Aggregation

Risk aggregation is the method or process by which individual risk may be combined for the purpose of reporting or treatment, or to obtain an integrated risk profile or risk score. Decisions regarding I&T risk management are more beneficial to the enterprise if risk is managed from the perspective of end-to-end aggregated risk. An aggregated view of risk supports complete and thorough review of risk appetite and risk tolerance, and always surpasses—in terms of enterprise benefit—relatively isolated recognition and/or treatment of risk.

I&T-related risk is often grouped together by risk type, similarity of risk response or specific control treatment. For example, if an enterprise access management approach generates repeated audit findings or control deficiencies across different business or mission areas, then an enterprise initiative in access management may resolve the issue.

The financial impact of risk is often aggregated, for executive or board-reporting purposes, into ranges of monetary loss that can be expected if certain types of risk are realized. Many enterprises maintain a set of impact criteria and risk tolerances expressed in financial terms. Risk aggregation and reporting are a current requirement for many financial institutions subject to the Basel Committee on Banking Supervision (Basel Committee) supervisory process.[25] This requirement is driving a discussion between senior management (or their delegates), the risk management function/staff and the board of directors on what is the appropriate level of aggregation and quantification of risk that would be acceptable and helpful to the board of directors to make informed decisions..

The financial impact of risk is often aggregated, for executive or board-reporting purposes, into ranges of monetary loss that can be expected if certain types of risk are realized.

8.7 Risk Response Selection and Prioritization

The previous sections list risk response options. This section focuses on distinguishing, evaluating and selecting appropriate responses among those options, given a specific risk context. The following parameters need to be taken into account in this process:

- **Cost of the response**—In the case of risk transfer, consider the cost of the insurance premium; in the case of risk mitigation, consider the cost of implementing, maintaining and testing controls.
- **Importance of the risk addressed by the response**—Consider priority or rank on the risk register.
- **Capability to implement and maintain the response over time**—The more mature an enterprise is in its risk management capability, the better the responses that can be implemented; when the enterprise is rather immature, some very basic responses may be used and improved over time.
- **Effectiveness of the response**—Consider the extent to which response activities will reduce the frequency or impact of the risk, should it materialize.
- **Other I&T-related investments**—Investing in risk response measures always competes with other I&T-related investments, and requires careful deliberation.
- **Other responses**—One response may address several risk types while another may not; risk may be aggregated and subsequently addressed with a common response.

Sometimes the effort or resources required for responses (e.g., the collection of controls that need to be implemented or strengthened) will exceed the available capability of the enterprise. In this case, decisions on prioritization, organizational skill and expertise are required. Possible risk response options can be grouped as follows:

- **Quick wins**—Quick wins include very short-term, time-efficient and effective responses to high-impact risk.
- **Compliance obligations for which there is a non-negotiable requirement**—Managing the risk of noncompliance should be done in conjunction with other risk responses to avoid duplicative or overlapping work.[26]
- **Business case required**—More expensive or difficult responses to high-impact risk require careful analysis and management decisions prior to investment. Responses in this category may also include outsourcing the management of risk that the enterprise cannot address internally.
- **Deferring and/or continued monitoring of conditions**—Enterprises may defer the response and continue monitoring to determine if changes to the identified risk or environment warrant a different response.

[25] *Op cit* Basel Committee on Banking Supervision
[26] See also Section 8.6 Risk Aggregation in this publication.

Figure 8.1 traces the broad process for selecting and prioritizing risk responses.

Figure 8.1—Risk Response Selection and Prioritization

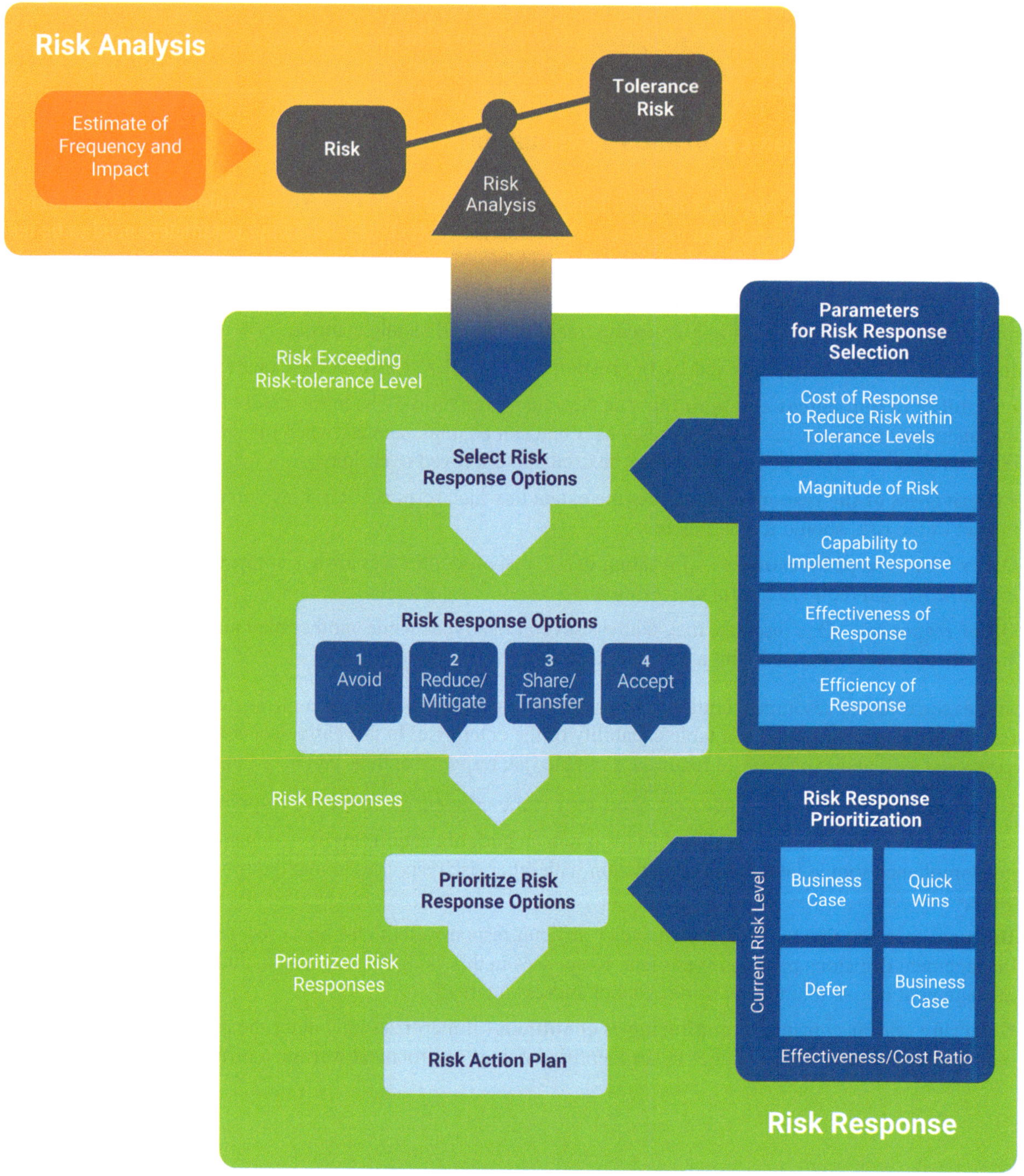

9781604208191